PHILOSOPHICAL NOTES

THE EVOLUTION OF EVOLUTION

St. George Jackson Mivart

Introduction by Gary Furnell

This edition published in 2023 by Connor Court Publishing Pty Ltd

First Published in the American Catholic Quarterly Review, VOL. XX , 1895 —
No. 80.

Connor Court Publishing Pty Ltd
PO Box 7257
Redland Bay QLD 4165
sales@connorcourt.com
Phone 0497-900-685

Printed in Australia

ISBN 9781922815538

Cover design: Gaetano Lauti-Paluzzo

Cover Picture: St. George Jackson Mivart. Photograph by Barraud & Jerrard. Wellcome V0026864.jpg. Wikipedia Commons.

Philosophical Notes

Asking questions on philosophical questions

INTRODUCTION

Mivart's essay on evolution rewards the meditative reader. First, Mivart provides a brief but cogent outline of the changing circumstances and variations in thought that *over centuries* facilitated the development of naturalism and atheistic evolution. This sort of background is particularly useful because each generation, enmeshed in the peculiar conditions of their own time, often neglects to explore the past cultural twists, oversights, discoveries and aberrations that led to today's assumptions and outlook. Then Mivart highlights the internal contradictions that fracture naturalism and he indicates some of the problems in the evidence used to support atheistic evolution. These fractures and variations of these evidential problems remain today.

St George Jackson Mivart (1827—1900) was an esteemed British naturalist, a Fellow of the Royal Society; vice-president (twice) of the Zoological Society and a Fellow of the Linnean Society. He was an early advocate of Darwinian evolution but parted company as its prominent proponents championed an atheistic expression of evolution. He thought *theistic* evolution made better sense of all the evidence. Mivart was an ardent—some would say an argumentative—Catholic.

He wrote that atheistic evolution was a sort of semantic oversight: evolution from what? It posited a process but neglected to explain the origin of the material exploited by the process. Was matter eternal, or did it inexplicably pop into existence from nothing? Theistic evolution had a positive and sufficient answer to this question and so—with its starting point established— could properly describe itself as an evolution.

Mivart argued for the uniformity of natural causes in an *open* system and he opposed the contrary presupposition: uniformity of natural causes in a *closed* system. He believed that reality could not be explained solely by the energies, entities and processes already existent. Rather, the energies, entities and processes had a divine origin. Put simply, Mivart maintained that the Ultimate was Personal, not impersonal.

The nature of this Person is not addressed in the essay. Mivart argues for theistic evolution without promoting his own religion. His aim was to highlight the logical and evidential superiority of theistic evolution over atheist evolution.

He enjoyed the irony that atheist evolutionists continued their rational exploration of the universe while announcing that everything that existed was irrational: a product of time plus chance. Theistic evolutionists do not have this problem, believing a reasonable God made a reasonable (although largely mysterious) universe. Man's reason, properly used, had its place—albeit limited by our common finitude.

Mivart lectured in anatomy and saw biological patterns—obvious everywhere in nature including human biology—that resisted attempts at explanation along atheist evolutionary lines. Linnean catalogues, for example, were only possible because of these stable patterns.

Mivart argued that atheistic evolutionists had taken a helpful biological insight—natural selection—and through their version of evolution had made it into an all-encompassing ideology. Today, this imperialistic project continues. Spheres as diverse as economics, sociology, medicine, politics, sexuality, psychology, psychiatrics, linguistics, religion and ritual—every aspect of life in fact—are routinely subjected to evolutionary analysis. Theistic evolution escapes this presumption because it doesn't need to explain everything through physical evolutionary processes alone.

Instinct, Mivart wrote, was an especially profound and precise facility whose origins found satisfactory explanation by theistic evolution but defied attempts at satisfactory explanation by atheistic evolution. Instinct works to incredible effect in the humblest of organisms. Mivart provided the example of injured crabs that break off a limb—so regrowth could occur—precisely where the limb was damaged. Mivart could have easily multiplied examples. I would add the existence of the human conscience—stable and universal in its nature, enigmatic in its activity—is likewise better explained by theistic as opposed to atheistic evolution.

Mivart thought atheistic evolution was doomed it to a speedy obsolescence because of its inconsistencies and the contrary evidences. In this, he was wrong. When his essay was first published 140 years ago, he doubted if atheistic evolution would last longer than a few more years. He was confident that theistic evolution would supplant it.

Although he doesn't use these terms, Mivart suggests that atheistic evolution is the favourite child of a favoured paradigm: the Enlightenment's truncating scientism. Predicting atheistic evolution's doom, he may have been better to say: When the paradigm changes—and this might be sudden—the favoured child might be abandoned.

It seems the paradigm is changing. In many academic faculties the Enlightenment's central tenets—triumphant human reason; the unequivocal benefits of science and technology; an end to human faults and follies through improved social conditions, health care and education— are being portrayed as power plays that justify their own activities, with many of these activities consequently exposed (correctly or incorrectly) as exploitative, destructive of the biosphere or aggrandising.

If the Enlightenment is successfully dismantled, it's likely Darwinism won't dominate as it does now. Theistic evolution may not dominate either. It's possible *any* idea might emerge and capture the imagination of restive people.

Darwinism is being dismantled, but not by theists: by biochemists amazed by the abundant cellular evidence of intricate, deliberate design. If he were alive today, Mivart might populate his essay with examples of fastidious design coming from molecular biology. Remarkable nano-machines, all sensitive to their minute cellular environment, abound. Prime-movers hauling bundles of chemical energy along self-constructing carriage ways, turbines, drives with cams and universal joints, protein factories, propelling engines and more—all of staggering complexity—are being brought to our astonished apprehension. Stunned wonder is one appropriate response.

Two hundred years before Mivart, Pascal wrote:

Let man then contemplate the whole of

nature in her full and lofty majesty, let him turn his gaze away from the lowly objects around him; let him behold the dazzling light set like an eternal lamp to light up the universe, let him see the earth as a mere speck compared to the vast orbit described by this star, and let him marvel at finding this vast orbit itself to be no more than the tiniest point compared to that described by the stars revolving in the firmament. But if our eyes stop there, let our imagination proceed further; it will grow weary of conceiving things before nature tires of producing them. The whole visible world is only an imperceptible dot in nature's ample bosom. No idea comes near it; it is no good inflating our conceptions beyond imaginable space, we only bring forth atoms compared to the reality of things. Nature is an infinite sphere whose centre is everywhere and circumference is nowhere. In short it is the greatest perceptible mark of God's omnipotence that our imagination should lose itself in that thought.

Gary Furnell is the author of *The hardest path is the easiest: exploring the Wisdom Literature with Pascal, Burke, Kierkegaard and Chesterton*, (Connor Court, 2021).

THE EVOLUTION OF EVOLUTION

St. George Jackson Mivart

St. George Jackson Mivart,
(Wikipedia Commons)

WE have had many treatises on evolution —the evolution of species, of the solar system, of the sidereal universe, of law, of art, of religious worship, of political organization, of systems of theology, etc., but as yet, so far as the present writer knows, no one has expressly considered "the evolution of evolution itself."

By this phrase we mean the origin, development, present state and probable future of that system of thought with which the teachings of such men as Herbert Spencer, Darwin, Huxley, Tyndall, Karl Pearson and Professor Haeckel have made us all more or less familiar.

The main characteristic of this system is its persistent endeavor to explain all higher energies in terms of lower activities, and so, while denying the possibility of creation without pre-existing material to affirm the universal origin of all things, with all their powers, from others which avowedly have no such powers to bestow. A second not less noteworthy character is its neglect to seek for or acquire any solid intellectual ground for its own support; so that while it, in fact, bases every power of the universe on impotence, it derives all the highest faculties of the intellect from nescience[1], and practically proclaims Unreason Lord of the Universe!

Strange, that a system so essentially irrational should have been slowly built up through the efforts, or with the approval, of so many gifted minds, to replace another which once commanded the allegiance of the keenest intellects the world has known!

That latter system possessed, indeed, the essential power to retain mens' allegiance on to our own

---

1   nescience: lack of knowledge or awareness

day, though the effect of personal deficiencies on a succession of events and changed circumstances (which alone could not have caused such an overthrow), made the continuance of its sway impossible.

It seems to us that few investigations could be more interesting than a thorough study of the decay and overthrow of the scholastic philosophy, together with the uprising, in its place, of that modern system which Mr. Arthur Balfour[2] has lately termed "naturalism", with its attendant net-work of skepticism, in the meshes of which Mr. Balfour himself remains helplessly entangled.

A few suggestions towards such a study are what we purpose here and now to offer to our readers, with the addition of some notes as to a few scientific facts which seem to indicate that "evolution", in the popular sense of that term, is a theory which has had its day.

If it is strange, as we have said, that such an incoherent system should have replaced one so

---

2   Arthur Balfour, former conservative Prime Minister of Great
    Britain 1902 to 1905.

solid and well laid-together, it is more strange still that evolutionists should regard that replacement with the satisfaction they do. It is as if some graceful mediaeval fane, with its slender soaring columns, its traceried windows, its dexterously arranged system of thrusts, lightly but solidly suspending in mid-air a fair groined roof, had fallen down in ruin, and that men having roughly framed an imperfect shelter from its fragments, should chuckle and congratulate themselves thereon, as on a positive architectural progress and improvement! Yet some of our readers may very naturally object that the intellect is not free, but must follow evidence, and, therefore that the earlier system could not have been so excellent as we suppose, since otherwise it must have continued, willy nilly, to command the assent of all, or almost all, capable men.

It is true that the intellect, in the abstract, must follow evidence; but men are by no means all intellect. They are largely dominated by their emotions, and many of them are readily carried away by each successive fashion of the day. Moreover, it is not everyone who adheres to a true system who can skilfully show forth its claims upon the assent of others; nor does it even follow that every skilful advocate who is faithful to it as a whole, is sound in every one of his views regarding it. Moreover, there are such defects as sloth, self-seeking, undue subservience to personal influence and neglect of fidelity to old truths under new circumstances.

Certainly the change from the intellectual harmony and completeness of the thirteenth century to the discord and disarray of the sixteenth, is wonderful. It was a transformation, indeed, affecting the whole of life, and most potent in its consequences. But this change may be paralleled by that which has transformed the domination of the Church as it existed under Innocent III, into the feebleness of the days of Pius VI.[3]

Under the former Pontificate it seemed that ecclesiastical authority could never lose the control which it then possessed over every department of national and social life. In the affairs of every nation, of every city and township, and, indeed, of every family in every land of Christendom, the Church intervened through her sacramental ordinances,

---

3   Pius VI, Pope from 1775 to his death in 1799.

her stately ceremonial and her supreme authority. Yet after a very few years the ignominy to which Boniface VIII[4] was put by the emissaries of Phillipe le Bel[5], led to his death, while in that pontiff's person the Church underwent a humiliation from which she has never recovered.

This fact is not, of course, to be explained by any real weakening of the Church's spiritual weapons, but by the sloth, self-seeking, undue subservience to personal influence and neglect of fidelity to old truths under new circumstances on the part of some or other of her ministers. Had the bishops of France been staunch in their loyalty to Boniface, had the inferior clergy been ready unhesitatingly to support their bishops in upholding the Pope against the king, in spite of royal violence and persecution and in adherence to that most logical Bull known as "Unam sanctam"[6], the humiliation of the Church would have been delayed for at least

---

4 Boniface VIII, Pope from 1294 to his death in 1303.
5 Philip IV, called Philip the Fair, was King of France from 1285 to 1314.
6 Unam sanctam – issued by Pope Boniface VIII on 18 November 1302 it stated the propositions on the unity of the Catholic Church, the position of the Pope as supreme head of the Church, and submission to the authority of the Pope in order to belong to the Church.

another generation. Changes due to the increase of knowledge, culture and refinement would, of course, have inevitably taken place, but we see no reason to think that these could not have been welcomed and provided for without injury to the hierarchical constitution of Christendom, nor that churchmen need have let power slip from their hands as they did.

A parallel case is presented by the overthrow of that metaphysical system known as the Scholastic Philosophy. We see no reason why the inevitable changes in the intellectual order which advancing physical and mental science necessitated, need have caused any philosophical breach of continuity. But such a breach once effected, the circumstances of the age could not but rapidly widen it with fatal efficacy; for never since the origin of the historic period did the leading races of mankind simultaneously experience such a rapid succession of religious, political, intellectual and physical changes as between the end of the fifteenth and the middle of the seventeenth centuries.

The breach itself must be admitted, however reluctantly, to have been largely due to the teachers of the "old learning" in the latter days of its general acceptance. An unwise conservatism and an invincible repugnance to the labor of welding together new truths with old ones, made them neglect those physical sciences which were then freshly budding forth, and rely much on seductive reasoning and great subtlety in verbal distinctions, as if the sphere of the verbum mentale could be substantially and greatly enlarged by playing cunning tricks with the verbum oris.[7]

Centuries before, that wonderful man, Roger Bacon, had warned the philosophers of his day of their danger, and wisely counselled them to have recourse to experiments and useful observations of facts, instead of relying so largely upon mere processes of ratiocination. In the sixteenth century it was almost too late to arrest the philosophic decay. Fresh discoveries and strikingly novel observations were accumulating on all sides. A new world had been disclosed by Columbus, with races of men and wonderful civilizations, which had remained for centuries unknown, while birds, beasts and reptiles utterly foreign to anything

7   Latin – word of mouth.

before known were being again and again brought over to Europe. But another discovery was even more important in transforming the intellectual and moral condition of Western Europe. Through the fall of Constantinople men became acquainted with the literature of ancient Greece, and there arose a violent passion for it — a passion which diffused and revived Platonism and prepossessions hostile to the whole Christian system.

A flood of moral corruption deluged Italy and spread to France and Germany, while a contempt for the ideas of the men of the preceding generation — men who had known neither the Transatlantic nor the Hellenic worlds — whose philosophical system must, therefore, they thought, be as inapt and barbarous in its conceptions as the latinity whereby those conceptions were expressed.

Thus, by the combined effects of neglect on the part of professors of the " old learning", and of impatient intolerance on the part of those of the "new", the first breach was effected in the philosophic continuity of Christendom, and the ground became thereby prepared for the germination, or "evolution", of the seeds of that hypothesis of evolution which has culminated before our eyes.

The rich harvest which physical science at once began to yield (as it always yields) to its enthusiastic cultivators very soon enormously widened — as we before hinted — the gulf between the old and the new systems. Yet the students of physics were perfectly right and fully justified in adopting methods and seizing upon truths which had, up to that time, been too much neglected, and it was but natural that success should quickly and greatly augment their enthusiasm. But, unfortunately, while thus seizing upon truths which had been before neglected, they neglected truths — and most fundamental truths — which, up to that time, had been constantly recognized. Not that all, or even the most conspicuous, devotees of "the new learning" were scientifically successful. Absurd, indeed, has been the admiration and praise bestowed upon that morally meanest of mankind. Bacon, and his "Novum Organum"[8], for, so little did he recognize or allow the claims of the intellect, and so absurdly servile was he in his mere quest of facts, that his

---

8 Novum Organum (New organon), a philosophical work by
Francis Bacon, published in 1620.

efforts were as barren of results as those of any of the school men upon whom he poured out his very foolish scorn.

The universal and objective validity of those primary and fundamental intuitions which can alone afford a logical foundation for even the lowest physical science, being ignored or repudiated, the gates of human thought were thrown open to admit that long procession of ephemeral follies put forth by successive sophists from Descartes, Locke and Hume, though Kant, Hegel, Schilling and Hartmann, to the Spencers, Huxleys, Tyndalls and Herschels of our own day. Therewith the "evolution of evolution" has run its course and the "unreason" of men has enthusiastically proclaimed the supreme "unreason" of the universe

That philosophical heresiarch Descartes, the *fons et origo*[9] of the later follies of all the men above named, the great exemplar of all "topsyturveydom" and who based "intuition" on "ratiocination", at once aspired to a mechanical exploration of the universe. To him is due that supreme folly of our own day

---

9   Fons et origo, Latin term meaning "source and origin".

which teaches (as Professor Huxley has taught) that to let a man know his house is on fire is not the cause of that process of " running home" which he may thereupon set going! That well-meaning, puzzle-headed Englishman, Locke, by reducing all our sources of knowledge to "sensation" and "reflection", supplied everything that was necessary for the development of the "idealism" of Berkeley and his followers to the present day.

But "idealism" is only the other side of the shield of "materialism"; since for neither system is there any source of knowledge beyond "impressions" or "feelings" ; and our "sensations" are represented as being "impressions" or "feelings" in their most vivid form.

Therefore beyond these we can (according to these systems) know nothing, and though the piety of Bishop Berkeley led him to regard the universe about him as a phantasmagoria[10] played off by God upon his (Berkeley's) own mind, that acute mocker of his fellows, Hume, saw clearly that from such premises very different conclusions followed. He therefore

---

10 Phantasmagoria, a sequence of real or imaginary images like that seen in a dream.

amused himself, and gained the renown he greedily desired, by drawing forth those consequences and depicting them to the stupid amazement of his contemporaries who took him seriously. Hume saw, clearly enough, that if "impressions" or "feelings" are the ultimate source of knowledge and its only trustworthy elements, then not only is God an impossible object of knowledge, but no man can have any knowledge of himself — of the reality of his own existence.

Every "impression" every "feeling" is a single, individual subjective state which cannot tell us anything about the real state of our objective world — not even its existence. Still less possible is it for mere subjective "feelings" to tell us anything about "continuity", "succession", or "causation", or inform us either as to what, if anything, gives rise to "impressions" or what, if anything, experiences them. Yet, strange to say upon such a basis he gratified himself and proved the amazing gullibility of Carlyle's "fools"[11] by pretending to construct a positive system upon two magic words

---

11 Thomas Carlyle, Scottish historian (1795-1881) when asked about the population of England replied: "Thirty million, mostly fools".

— "association" and "custom". By "association" he signified the tendency of the individual to join together in imagination and expectation successive and simultaneous experiences. By "custom" he denotes a sort of solid or collective "association", by which masses of men came to acquire similar combined impressions and expectations.

Thus it was, he said, that our fleeting impressions are able to suggest to us a permanent and orderly world, and so (he taught) it comes about that our rapid stream of individual feeling, deludes us into a persuasion of our continued personal existence and that the complex impressions we denote by the terms "continuity", "succession" and "cause", have an existence in our consciousness.

It is really impossible to believe that a mind so acute as that of Hume, could have put forward such a system, save with "his tongue in his cheek" — "association" carried on by what cannot associate because it has no being, and "custom" experienced by a society which has no existence either as a whole or in its component elements — we having no more intuition of the existence of either than we have of the existence of God! Hume, in whatever corner of the universe he may now be, must surely enjoy — amongst other modes of motion — a self-complacent chuckle over Professor Huxley's little book about him[12] and the solemnity with which *jeux d' esprit*[13] are seriously commended to the admiration and imitation of the professor's contemporaries.

But "association" has played a very important part in the evolution of evolution. That feelings, sense-experiences, emotions and ideas which have existed simultaneously or in succession do tend to become associated is, of course, an unquestionable fact, and the English sensists (culminating in John Stuart Mill and Herbert Spencer), have made use thereof to explain the genesis of our ideas from sensations and to attempt to show how human self-conscious intellect may have been evolved from the psychical faculties of a lower animal. "Nihil in intellectu quod non prius fuerit in sensu"[14] is the motto of the whole school; and, taken in a certain sense, the dictum is good and true. For it is a necessary consequence of

---

12 Hume by Professor Huxley, London, Macmillan and Co, 1879.
13 jeux d'esprit, a work of literature displaying wit and clever-ness.
14 Latin, literal translation: There is nothing in the intellect which was not first in the sense of.

our combined bodily and mental constitution, that no idea can be present to our minds save by the aid of some mental image present to the imagination and we can imagine nothing of which we have not previously had some sensuous experience.

Hence the mistake made by those who think that no conception can be true unless it can be "mentally visualized" by us. This is, of course, really a great mistake, because it amounts to a denial of the distinctness of our ideas (such as those of "existence", "non-existence", "absolute being", "possible being", "truth", "necessity", etc.) from our feelings.

But though "feelings" and "mental images" are necessary antecedents and accompaniments of our ideas, they are so essentially distinct from the latter that the dictum "*Nihil in intellectu quod nunquam fuerit in sensu*"[15] is no less true than the former adage. These dicta are indeed complementary truths, the acceptance of both of which is necessary for a right understanding of human reason, which is intellect

---

15 Latin, literal translation: There is nothing in the intellect, in the sense that it has never been.

energizing in and subserved by a material, corporeal power.

Confusion between imagination and conception is one of the commonest of modern errors, and is one through which alone "the evolution of evolution" has been possible. A clear perception of the essential distinctness which exists between these two faculties is the first requisite for a sound psychology. But not only did this fatal confusion impair a due recognition of the claims of reason by men of the "impressionist" school, any such recognition was made impossible for them by their denial of

(1) our powers of apprehending universal and necessary truths, (2) of knowing with certainty our own continuous, and (3) generally of apprehending all objective, existence; it being assumed that human knowledge is rigidly confined to "sensations" and "sense-impresses", and is essentially "relative", and therefore necessarily divergent from the realities of things as they exist for any super-human intelligence.

Even Kant, who sought to affirm against Hume

the validity of pure reason, and by his categorical imperative seemed to assert, in the most forcible manner, the objectivity of ethics, really remained bound in the chains of subjectivism, since our highest and seemingly most certain perceptions were but the outcome of human, conditioning "forms of thought".

Sir William Hamilton[16] was long regarded as a successful opponent of skeptical and sensuous philosophy, and yet his doctrine of "the relativity of knowledge " opened the door to yet more vigorous and successful assertions of evolutionary mental philosophy, a most complete exposition of which is to be found in the pages of Mr. Herbert Spencer.[17]

The possibility of mental evolution (of intellect from sense) having been made first conceivable, then possible, and finally (in their own eyes) certain, by these successors of Locke, it only remained — in order that the modern doctrine of evolution should gain acceptance — to show that

---

16 William Hamilton, (1730 –1803), was a British diplomat (1754-1800), antiquarian, archaeologist and vulcanologist.
17 Herbert Spencer (1820-1903) British philosopher, biologist, anthropologist, and sociologist famous for his theory of social Darwinism.

corporeal evolution might be accepted also.

The progress of biological science during the eighteenth century had prepared men's minds, by the new views it had opened up, for yet more startling suggestions. Buffon (1707-1788)[18] was especially remarkable for the hypotheses he emitted about the origin of the earth, the nature of generation, and the relation between the animals of the old world and those of the new. Kant and Goethe[19], as well as Buffon, put forth ideas respecting the transformation of species, while Lamarck[20], in the very beginning of this century, distinctly promulgated the doctrine of the origin of new species from pre-existing ones through the influence of surrounding circumstances. His views gained small acceptance, and then fell into utter discredit, yet of late they have obtained many adherents, especially in the United States,

But the really influential conception, the emission of which was to act as a spark on a mass of stored-

---

18 Georges-Louis Leclerc, Comte de Buffon was a French naturalist, mathematician and cosmologist.
19 Johann Wolfgang von Goethe (1749-1832) was a German writer.
20 Jean-Baptiste Lamarck, (1744-1829) was a French naturalist.

up gunpowder, could never have been formed but for the excellent labors of successive naturalists, who little foresaw the future, startling outcome of their toils.

Thus it was Linnseus[21] who, in 1735 and 1758, set forth the first practical system of biological classification, and originated a convenient system of nomenclature without which it would be very difficult to acquire, and quite impossible to retain in mind, a knowledge of any very numerous set of objects. The careful definitions of Buffon's collaborator, Daubenton[22], and the untiring industry of John Hunter[23], accumulated treasures of anatomical and physiological knowledge; but it was Cuvier[24], above all, who, by his indefatigable labors in investigating the structure of animals, living and extinct, prepared the necessary materials for the theory which was to come. Not less important was the circumstance that Werner, Hutton, and others

---

[21] Carl Linnaeus, (1707-1788) Swedish botanist, zoologist, and physician who is known as the "father of modern taxonomy".

[22] Louis-Jean-Marie Daubenton (1716-1800) was a French naturalist.

[23] John Hunter FRS (1728-1793) was a Scottish surgeon.

[24] Georges Cuvier (1769-1832) was a French naturalist and zoologist.

had shown the earth's crust to consist partly of stratified and partly of unstratified rocks, and that William Smith[25] (the father of English geology) demonstrated the existence of definite and uniform relations between different strata and the fossils they respectively contained.

Subsequently, the labors of John Miller[26], in Germany, and Richard Owen[27], in England, were most fruitful of results, and by degrees it became known that:

1. Different geographical areas are inhabited by different organisms.

2. There is a relation and affinity between the past and present inhabitants of each such area.

3. There is a certain resemblance between fossil forms of more or less antiquity and some immature stages of existing living beings.

---

[25] William 'Strata' Smith (1769-1839) was an English geologist.
[26] John Miller (1715–c.1792), was a German engraver and botanist based in London.
[27] Richard Owen (1804-1892) was an English biologist.

4. There is often more or less similitude between the earlier stages of existence of many living animals and the adult condition of other living animals belonging to lower forms of life.

Evidently, these facts favored the notion that the latest forms of life had grown out of preceding ones, and the view that new species of animals and plants had been produced through the operation of some obscure and hitherto undiscovered laws, became widely accepted amongst the most eminent naturalists. Such a view, for example, was not only maintained by the late Sir Richard Owen, but he declared that the discovery of such a law was possibly the chief end which the best anatomists and physiologists had in view.

Some theory of "Evolution" was thus evidently soon to be "evolved". But what theory? The theory of evolution which has become so widespread and popular, and forms a part of what Mr. Balfour calls "naturalism", advocates an " evolution" of one kind. But there is another, and a very different theory of evolution, which we are persuaded is destined to supplant the former, and we believe there are abundant signs that it is destined to do so at no distant day. The popular form of evolution is non-theistic; that which we believe will, rather sooner than later, supplant it is Theistic evolution.

Any rational theory of evolution must consider the world as a whole. It must, therefore, include man, and take into account his higher no less than his lower faculties. It must account for what we regard as our perception of necessary truth; our self-conscious knowledge of our own continuous existence, and our perceptions of truth and goodness — of right and wrong.

For all men who are convinced they have such higher faculties, any theory of evolution which ignores them must be absurd. That reason, as we know it in consciousness, is the outcome of the mere play of physical forces, is a doctrine which has but to be stated to such men in order to be self-condemned. But to those who, as we have seen, ignore the essential characters of their own intellect, regard their ideas but as modifications of past sensations, and declare virtue to be but one form of pleasure; unreasoning evolution presents no such contradiction.

Without seeking to ascertain the reason why, we may be quite sure of the fact, that many men have greatly desired to be able to conceive of the universe as free from any Supreme Intelligence or All powerful will. Rather than admit the evident existence of such, they would, if they could, have passionately repudiated all evidence of design in nature, and eagerly proclaimed Unreason to be Lord of the Universe.

But so to do, was long impossible. The evidences of design in nature were too plainly to be read in the facts of animal structure; in the accord between organization and function; and in the marvellous phenomena of instinct. That the activities of organic life could ever be explained (as Descartes had suggested) by the mere motions of a cunningly contrived mechanism, was beyond all belief. Kant held it to be absurd even to think that any naturalist would ever arise who should be capable of explaining so much as the growth of a blade of grass mechanically.

But at last the hour of the cosmic clock struck; the man appeared; and soon, Darwin's hypothesis of the origin of species by "natural selection" was promulgated. Thereby, the advocates of mechanism obtained full satisfaction; the difficulties which beset the adorers of the god "Unreason", seemed to be removed, for not only mental but also corporeal transformation appeared most simply explicable. Such transformism, in its most popular and pernicious sense, was welcomed; and the world witnessed a new mental birth, "the evolution of evolution".

A mechanical explanation of nature, for those who desired it, really seemed to have become possible. By "natural selection" the most wonderful adaptation of structure, and the most divergent peculiarities of organization could, it appeared, all be explained by the conservation in the struggle for life of minute fortuitous variations transmitted to offspring. The situation became thus completely changed; the idea of "design" or "purpose" seemed at once to have become superfluous, and Haeckel coined a special word (Dysteleology[28]) to denote a science of aimlessness in nature.

The new hypothesis struck the most dangerous

---

28 Dysteleology is the philosophical view that existence has no
    final cause from purposeful design.

blow at Theism which any living man has witnessed, and its success was great. For "natural selection "seemed to make all so easy and obvious; and it soon grew, therefore, to be as attractive to the multitude" as its deification of mechanism and chance caused it to be enthusiastically welcomed by anti-theistic philosophers.

As to how it is now welcomed, and what is the present position of the popular theory of evolution, we will say a few words later on, but before doing so we desire to call attention to one very notable instance of its rejection.

When Darwin and Wallace promulgated their theory of "natural selection", the greatest of English anatomists and natural philosophers, Sir Richard Owen, did not welcome it. He was a believer in what he called the "ordained becoming" of new species by appointed "secondary laws", and was an advocate of rational, that is, theistic evolution. He declared' that species change "by virtue of inherent tendencies" thereto, affirming that such a succession of species by continuously operating law is not a blind operation, but rather intimates" a preconceived progress", and that such evolution, proceeding "towards a foreseen goal", shows in its "broad features" the unmistakable impress of Divine volition. He passed away from us with his mind unchanged in his views as to evolution and it can now be said that the greatest English comparative anatomist of this century has, after a consideration of the hypothesis for more than the duration of an entire generation, continuously and finally rejected it. This we believe is the greatest fact of our venerable anatomist's whole life and teaching. For this firmness and consistency he had for a time to endure the ordeal of disesteem and contumely from not a few anti-theistic men of science. But, as has been the case in previous instances, that which his contemporaries least appreciated may, we believe will, hereafter largely add to his fame, if it does not even constitute his greatest glory.

The opposite state of mind is most clearly exemplified by the noteworthy European naturalists, Professors Weismann and Haeckel. Their dicta will enable our readers to understand what is the present state of the Darwinian theory and to what a degree of absurd dogmatism that theory of evolution has been evolved.

That there is a Providence which "shapes our ends", rather that the universe is replete with purpose, are in the eyes of Professor Weismann errors so fundamental, that any asserted facts which imply them, are thereby, ipso facto, demonstrated not to be really facts, but fictions. That the cosmos is ruled by unreason seems to be an article *stantis val labentis ecclesiae.* Therefore when he finds that the effective action of "natural selection" in the origin of species, not only cannot be proved but that there are facts which positively conflict with it, the consequence is that "it is so much the worse for the facts". He tells us expressly that facts of this kind cannot really exist, since efficiency of "natural selection" is a truth demonstrated by a process of *reductio ad absurdum.*

Had we not absolute certainty thereof, we could not be sure that there is no design in the universe or deny the possibility that through the phenomena of nature "an increasing purpose runs".

He has said: "It is really very difficult to comprehend the process of natural selection in its details; and to this day it is impossible to demonstrate it in any one point... We accept it not because we are able to demonstrate the process in detail ... but simply because we must, because it is the only possible explanation that we can conceive. It alone can explain the adaptation of organisms without assuming the kelp of a principle of design ... it is inconceivable that there could be yet another capable of explaining the adaptation of organisms without assuming the help of a principle of design."

Professor Haeckel has recently delivered himself of what he calls a "confession of faith" which is even more noteworthy. His dogmatism and self-conceit combined, have afforded us the most amusing piece of reading that we have enjoyed for a very long time.

The following passage is an example of his readiness to lay down the law absolutely and at once, about matters concerning which the ablest physicists profess themselves to be tentative inquirers. He says "with regard to evolution, from primeval chaos to the present order of the cosmos".

"At the outset there is nothing in infinite space but mobile elastic ether and innumerable separate particles — the primitive atoms — scattered throughout it in the form of dust."

Those "primitive atoms" are no mysteries to him for he has evidently had them in his laboratory and overhauled them carefully.

Thus it is that he is able to describe them so fully to us as being of a definite magnitude and spherical in shape. He has ascertained — no doubt by a series of careful experiments — that they are "inelastic", "impenetrable" and "indivisible", and, *mirabile dictu*, "inert". Yet in spite of this inertia they have a tendency to unite in small, definite groups, which groups upon careful examination, he found, turned out to be those "identical atoms" which so few physicists before Haeckel had been able to isolate, still less to build up. These chemical atoms showed, of course, their special properties, and the Professor does not seem to have rested till he found out that these properties had a cause, that it was but one cause and what that one cause was. These atoms he declares to be "solely conditioned by the varying number and disposition of the similar primitive atoms of which they are composed". It is to be regretted that Professor Haeckel has not given any diagrams of the arrangement of these "primitive atoms" in, say, "an atom of carbon" or an "atom of oxygen".

Having thus penetrated into the most intimate constitution and surveyed the origin, from "the outset" of the inorganic world, we need not wonder that the mystery of the coming of life and of the first origin of organization and growth have been easily mastered by him. The history thereof he gives, somewhat too briefly, as follows:

> "After the glowing sphere of the earth has cooled down to a certain degree, drops of fluid water precipitate themselves on the hardened crust of its surface — the first primary condition of organic life. Carbon atoms begin their organism — engendering activity, and unite with the other elements into plasma combinations capable of growing. One small plasma-group oversteps the limits of cohesion and individual growth; it falls asunder into two similar halves. With this movement begin organic life and its most distinctive function, heredity."

This appears to have been listened to with admiration by the meeting of naturalists at

Altenburg to whom it was addressed. A profane person, however, has characterized it as one of the most barefaced efforts ever attempted to disguise hopeless ignorance by empty verbal combinations. But even such wonderful discoveries as the ultimate constitution of matter and the origin of life and organization by no means satisfy our professor, whose motto should certainly be "Excelsior"! "From this successfully scaled height of knowledge", he triumphantly exclaims, "there open up before our joyously quickened spirit of research and discovery, new and surprising prospects, which promise to bring us still nearer to the solution of the one great riddle of the world".

Nothing remains hidden from the gaze of so physically inspired a prophet. Like a Neo-Platonist of Alexandria he has, while yet living, attained to a direct and immediate vision of the deity, manifested, however, in a somewhat peculiar shrine, namely, under the bell jar of an air pump. The light therein visible is a sort of Shekinah[29], for he tells that in that "it is the vibrating ether we see", while the

---

29 Shekhinah is the English transliteration of a Hebrew word meaning dwelling or settling of the divine presence of God.

ether is nothing less than "God the creator, always in motion". We can regard it — the cosmic ether — (he proclaims) as all comprehending divinity, and upon this found the thesis, "Belief in God is reconcilable with science".

Let us now turn to briefly note the essential characteristics of the rational, theistic theory of evolution, and especially note the differences which exist between its conception of the inner nature and properties of the organic world and that of the Darwinian conception. According to the theistic conception of evolution it is the outcome and manifestation of powers, principles and laws impressed on the material universe in the first instant of its creation. If species change they change by virtue of inherent tendencies, their conception being the manifestation of a preconceived progress which takes place not blindly or by chance, but by continuously operating law towards a foreseen goal.

The Darwinian view, the view of "naturalism", is very different. According to it (and this is its greatest fault) no inherent tendencies, no innate laws, govern either individual development or the evolution of new species. All is due to the chance action of small congenital variations, each such variation being itself accidental, the whole beauty and order of organic life, though, of course, determined by invariable unconditional antecedents, being but the merest outcome of the merest chance.

If such is really the case, then living organisms offer indeed a singular contrast to non-living inorganic nature. There at least we perceive that every so-called element, every chemical compound, every crystal and every non-crystalline inorganic body has its own innate powers and properties, and is subject to specific laws from which it never deviates. Such bodies act in many ways on one another, but in every such action the reaction of the body acted on takes place strictly according to its own innate laws and endowments.

Among the most recent valuable works of science which controvert this system and give good evidence against it, is one by Mr. William Bateson. He strongly supports that view which would forbid us to regard the world of living beings as any less governed by innate laws than is the inorganic world.

Granting, for argument's sake, that new species have arisen through variation, his work is devoted to examining the question whether such variations are indefinite and minute, so that if the whole series of them could be seen they would appear "continuous", or whether such a view of them would imply sudden and considerable changes — so that the variations would seem evidently "discontinuous".

This discontinuity may be as plainly and unmistakably manifested in the most minute structures as in all large ones, and Mr. Bateson's volume is principally occupied about cases of bodily symmetry whether normal or abnormal.

One minor form of symmetry is that which determines the patterns which may be formed, whether upon large surfaces of large organisms, or on the most minute structures. It certainly does appear to us that the evolution of many of these patterns constitutes a difficulty for naturalism which is none the less great because it has been so generally overlooked.

As to this Mr. Bateson says:

> "If anyone will take into his hand some complex piece of living structure, a passion flower, a peacock's feather, a cockle-shell, or the like, and will ask himself how that came to be so, the part of the answer that he will find it hardest to give, is that which relates to the perfection of its pattern. And it is not only in these large and tangible structures that this question arises, for the same challenge is presented in the most minute and seemingly trifling details. In the skeleton of a diatom or of a radiolarian, the scale of a butterfly, the sculpture on a pollen-grain or an egg-shell, in the wreaths and stars of nuclear division, such patterns again and again recur, and again and again the question of their significance goes unanswered. There are many suggestions, some plausible enough, as to why the tail of a peacock is gaudy, why the coat of a pollen-grain should be rough, and so forth, but the significance of pattern is untouched by these. Nevertheless, repetitions arranged in pattern exist throughout organized nature, in creatures that move and in those that are

fixed, in the great and in the small, in the seen and in the hidden, within and without, as a property or attribute of life, scarcely less universal than the function of respiration or metabolism itself."

One of the most obvious characters presented by our body, and by the bodies of all the animals most familiar to us, is that each has a right and a left side, and that these two sides, and their parts, correspond, as our right hand proverbially resembles our left one. When deeply considered, this fact is by itself sufficient to prove that the body of an animal has its own innate laws which regulate its development, for this kind of correspondence — technically called "bilateral symmetry" — shows itself not only in these familiar conditions, but in the results of disease, and in very peculiar structures found in exceptional animals of special kinds. Indeed, on the hypothesis that a blood-relationship of descent binds together different animals, nature actually forces upon us the perception that new and more intense forms of bilateral symmetry have arisen in comparatively recent geological time.

Thus naturalists now generally agree that birds have descended from reptiles, but the very diverse bilateral symmetry which exists between the two wings of birds for the one part, and between their two legs on the other part, is far more striking than any which is to be found in their hypothetical progenitors.

Mr. Bateson supplies us with numerous instances of similar and simultaneous bilateral variations.

Another form of symmetry is known as "serial symmetry". Such symmetry is most plainly seen in the successively similar segments and pairs of limbs in the centipede and its allies, but it is also to be traced in the structure of the human chest, with its successive ribs, and the series of bones (called vertebra), which comprise our spinal column, or backbone, and in the resemblances which can be traced between the arm and the leg, and between the hand and the foot.

Mr. Bateson in his third chapter gives many examples of sudden variation in parts which are serially symmetrical.

In man and beasts the bones (vertebrae) which exist in the neck are normally seven; and this is so whether the neck is enormously elongated, as in the giraffe, or exceptionally short, as in the whale or mole. The only exceptions to this rule are the manatee, which has but six, and the sloth, in which there may be as few as six, or as many as ten cervical vertebrae. In the human subject the last bone of the neck, which normally is entirely devoid of movable ribs, sometimes possesses them. Fifty-seven such cases are cited by Mr. Bateson, and it is interesting to note that in forty-two of these instances the ribs were present on both sides.

Variations are sometimes found in the number of vertebrate existing in successive regions of the spinal column, there being sometimes, for example, thirteen instead of twelve in the chest; four, five or six in the lumbar region, etc. To ascertain the exact correspondence between bones which thus differ in different individuals and species, has been a curious subject of inquiry among naturalists. Similarly, there are singular divergences, with occasional individual variations, in the number of bones which make up the wrist or ankle of different animals, and much ingenuity has been expended in trying to determine what are the precise correspondences in different cases of this kind. Mr. Bateson (we think with much reason), regards this quest, in the way it is often pursued, as but a vain one. Though nature's methods are simple, he urges, yet her simplicity is not ours. We are too apt in this matter to run into anthropomorphism, and allow ourselves to "fancy that nature has produced the forms of life from each other in the ways which we should have used if we had been asked to do it. If a man were asked to make a wax model of the skeleton of one animal from a wax model of the skeleton of another, he would, perhaps, set about it by making small additions to and subtractions from its several parts; but the natural process differs in one great essential from this, for in nature the body of one individual has never been the body of its parent, and is not formed by a plastic operation from it; but the new body is made again new from the beginning, just as if the model had gone back into a melting pot before the wax model was begun."

The author just cited also describes a multitude of instances of discontinuous variations in animals of the most varied and diverse classes. These variations have been detected in both external and internal bodily structures of the most varied kinds. They relate to variations of dentition, to the number and situation of mammary glands, to peculiarities of arterial distribution and modifications in the ducts of important glands — such as the kidneys, to the eyes of insects and shellfish, to insects' wings, the bones of ruminants, and especially in peculiarities of the digits (fingers and toes) of many animals.

The facts described by Mr. Bateson seem to us sufficient to prove the very frequent occurrence of discontinuous variation. To the consequent probability that new species have been evolved by the help of such, it has been objected, notably by Dr. A. R. Wallace, that his ample catalogue is a catalogue of monstrosities. Every one, as has been urged, knew that monstrosities from time to time occurred, but a new species — necessarily a symmetrical and well-organized form — could never have owed its origin to a mere monstrosity, such as a cat with two heads, a beast with superfluous or deficient digits, with half, or with double, the number of its proper supply of teeth, etc.

This criticism, however, is both unfair and exceedingly shallow. It is unfair because true as it is that we all know of the occurrence of monstrosities, very few of us know of the great frequency and enormous numbers of such variations as those Mr. Bateson notices. Besides this, any candid peruser of the book criticized must see that its author is perfectly well aware of the monstrous character of many of the varieties described by him, and has no intention of presenting such forms as the probable, or even possible, origin of new species. It is still further unfair, because a certain number of the variations described are perfectly harmonious and symmetrical changes, and can no more be said to be "monstrosities" than a tiger can be said to be a "monster" because it has not the mane of a lion, or a lion a "monster" because it has not the stripes of a tiger.

The real nature of many of the variations recorded by Mr. Bateson which may be said to be "monstrosities"? They are orderly and perfect structures in themselves, and by that very fact show that the organic world is not that inert mass of matter devoid of innate law Darwinism supposes, but is orderly in even its very aberrations from normal type. Mr. Bateson, among such instances, describes and figures a variation which has been observed in a saw-fly, in which the end of an antenna had taken on the form of a perfectly well-formed foot; also a beetle, of the genus Carabus, in which one limb was replaced by a pair of legs, each of which was perfect and normal in the details of its structure, with a number of other similar instances. In all these cases the details of the structures are orderly and quite different from the indefinite, fortuitous phenomena which we should expect to find in changes produced by external influences on bodies not capable of self- regulation by internal laws.

But among the mass of instances described there are some which are altogether harmonious and symmetrical discontinuous variations.

Yet, did we know of one case which was truly and unmistakably of that kind, that alone would suffice to make it probable that if new species arise by variation at all, such discontinuous variation is the kind of variation which has been efficient in their production.

One change of the kind, is that which occasionally

so affects the feathers of birds, such, e.g., as the moorhen, as to make them resemble the plumage of the apteryx and the cassowary. Such a variation amongst poultry produces the forms known as the "silky-fowls", sometimes called "emu-fowls", which are capable of perpetuation by breeding.

The long-haired varieties of goats, cats, and rabbits, are familiar enough; but it is less generally known that a similar variation exists in what are called "Peruvian guinea-pigs". Mr, Bateson mentions the capture of a common mouse with long black, silk-like hair, which, he adds, is specially "interesting, as showing that such a total variation may occur as a definite phenomenon without selection."

On the other hand, varieties have occurred which were entirely naked, but had a wrinkled condition of skin, and they produced young which were similar to their parents.

Now, in south Africa two creatures belonging to the same natural order as the mouse, are of about the size of a mouse, and have a burrowing habit. They are normally and naturally naked, with a wrinkled skin. If new species have arisen by variation, why should not these African animals have arisen by a sudden discontinuous variation, such as that, we know, may occur in the case of the common mouse?

Groups of plants are often characterized by having the parts which make up the flower in fives, or fours, or threes, as the case may be.

The tulips belong to a group having the parts or organs of its flower in multiples of three. Mr. Bateson gives an interesting example of a tulip having all the parts of its flower in fours. This variation, as he says, "is a large and decided one; but, it is more than this; it is not only large, it is complete. The resulting form possesses the character of division into four no less completely and perfectly than its parent possessed the character of division into three. The change from three to four is thus perfected: from the form with perfect division into three is sprung a form of perfect division into four. This is a case of a total ox perfect variation." Obviously, the perfection and symmetry of this remarkable tulip could have been in no way due to "natural selection". Why, then, should the tulips with their parts in three have owed that condition to "natural selection"?

One of the most curious parts of animal structure is the difference which exists between the build of all the American apes on the one hand, and those of the old world on the other. One of these differences is that between the number of bicuspid molars — grinding teeth — which have milk predecessors. In man, and all monkeys of the old world, there are only two on either side of each jaw, but in all the monkeys of the new world there are three such.

Mr. Bateson describes and figures the skull of an American spider monkey, in which there are four such teeth on each side of the upper jaw. All these are equally well formed, so that it is impossible to say that any one of them is supernumerary rather than another. Thus, it is demonstrated that a new and perfect form of dentition — an emphatically discontinuous variation — may suddenly arise.

Another most striking divergence of structure is that which exists between beasts (such as oxen, goats, deer, llamas, swine, and the hippopotamus), in which the number of toes used in locomotion is even — two or four — and those of other beasts {such as horses, asses, tapirs, and rhinoceroses), in which the number is odd — one or three.

In the even-toed group, the line of symmetry passes down in the middle between the two toes, while in the odd-toed group it passes down along the middle of the single toe (horse and ass) or along the middle of the middle toe of three (tapir and rhinoceros).

But a very exceptional variation is noted by Mr. Bateson. It is that found in the foot of a horse, in which, instead of a single digit there are two perfectly complete ones each bearing a hoof, while the line of symmetry tapers down between them.

Altogether the phenomena recorded in Mr. Bateson's book seems to us amply sufficient, by themselves alone, to deal a death blow to the otherwise absurd and childish theory that species owe their origin to a process of "natural selection".

But though natural selection is the crown and culmination of the "naturalistic" — non-theistic theory of evolution, we think it desirable to also say a few words respecting Mr. Darwin's ancillary theory of "sexual selection".

According to that notion all the special

characteristics of the male sex in each species —
all that to us seems beautiful, bizarre or revolting
(strength and nimbleness apart) — have been
evolved by means of the constantly recurring
exercise of the female, of her power of choice,
amongst contending suitors determined by
aesthetic considerations only.

How largely mere fancy may color not only the
readings but the very observations of some persons
appears to us to be remarkably well shown by
an account given by a Mr. and Mrs. Peckham,
of their observations respecting "sexual selection
in spiders".

They describe the male as courting the female by
dancing, during which he displayed to her his
attractive first pair of legs, of a delicate green fringed
with white, in various odd positions. The effect was
remarkable, for we are told she "eyed him intently",
appearing much "interested in his performance"
and finally encouraged him by "gazing in a softer
mood"; surely these spider's "sheep's eyes" are very
noteworthy.

Yet his apparent success could not free him from all

anxiety, as she would sometimes "make a sudden dash at him", a process which (seeing that the female is always larger and more powerful than her wooer), had a suspicious flavor of cannibalism about it. Indeed the female spider, "*Etiam in amoribus serva*"[30] is but too apt to bite and suck dry even her accepted spouse. Surely a very remarkable "post-nuptial" settlement.

We find it simply impossible to believe that the brilliant local coloration which certain old-world apes possess could have been due to aesthetic sentiments on the part of female apes and baboons, when we call to mind what are their psychical characteristics and what the physical powers of their would be spouses.

To produce such effects, the females of each species must have a taste in color as different as constant, while a multitude of remarkable or brilliant phenomena of color are found in parts which can never meet the eye during life.

One of the oddest of the many odd notions set

---

30 female spider kills and eats a male before, during, or after copulation.

afloat was that promulgated by Mr. Darwin himself — the notion, namely, that it has been feminine influence which has by degrees, as century succeeded century, gradually denuded our backs of the hair with which it was at first so copiously clothed. Now it is evident that the primitive ladies of the Kalmuck[31] and Persian nationalities differed widely in their sentiments as regards the beard; but nevertheless (if the theory is true), the females of every tribe and nation of mankind — in spite of the frequent mutations of fashion — must have unanimously and persistently agreed in abhorring hirsute shoulders, and this though their immediate pithecoid ancestors[32] entertained a directly opposite sentiment.

But the absurdity of "sexual selection" is now generally recognized and that of "natural selection" "as the cause" of the origin of species, is being recognized more and more widely. The work of Mr. Bateson and another one by Mr. Frank E. Beddard, F.R.S.[33], are strong indications that the tide is turning as regards non-theistic evolution, that it

---

31 The Kalmyks are a Mongol subgroup in Russia.
32 pithecoid an anthropoid ape.
33 Frank E. Beddard (1858-1925) an English zoologist.

has culminated and the process of descent begun. Great has been the part played in the evolution of evolution by the cell-theory — a theory of great value when not pushed to the length it has been. Mr. C. O. Whitman[34] of the United States now deliberately maintains and with very solid arguments, "the inadequacy of the whole cell theory" and has shown (as we long ago affirmed), that the difference between organisms formed of one cell and others formed of many cells, has been enormously exaggerated.

The popular Darwinian doctrines that the coloration of animals is due

(1) to their resemblance to their environment, or

(2) to their resemblance to some other animal dreaded on account of its dangerous properties, or

(3) its nauseous taste, are all carefully considered and satisfactorily disposed of by Mr. Beddard.

But a belief in non-theistic evolution can never be satisfactorily banished by physical science alone. To

---

34 Charles Otis Whitman (1858-1925) an American zoologist, one of the founders of classical ethology.

effect that, it is necessary to revive and disseminate far and wide some philosophical conceptions, especially their bearing upon the question as to the origin of species.

Considerable or small gaps between the various kinds of living creatures are manifest on all sides. The existing creation is plainly discontinuous. The facts as to continuity and discontinuity of variations, of successive forms of life and of surrounding conditions, are matters which demand most careful investigation, and investigations of the kind, such as have here been noticed, demand our gratitude.

But what hope is there of any student of nature arriving at the truth as to the continuity or discontinuity of specific origins, if he has not a clear comprehension of the great facts of discontinuity which are, on all sides, open to his mental gaze?

There is in the first place, the chasm which exists between every thing which lives and all that is devoid of life. Granting that the universe may have been so formed that on the occurrence of the preordained conditions life, previously created *in potentia*, should suddenly manifest itself "in act", that does not in the least invalidate the deep significance of the fact that for all our experience no life arises save from what already lives.

Secondly, there is the chasm between everything which feels and all that is devoid of sensation. Every one must admit that this chasm exists — every one who is not prepared to affirm that his paper and his pen feel each other.

Far greater, however, is the chasm which exists between every being capable of self-consciousness and a knowledge (however imperfect) of truth, goodness, and beauty, and every being devoid of self-conscious life.

The advocacy of these great truths is, at last, beginning to be patiently listened to, which is no small gain; and a rational physical kind of agnosticism will ere long take the place of that system which professes to understand science, but to have no knowledge of truths without which all science is absolutely, logically impossible.

The true nature of the organic world, the innate

properties of what we call "elements", and the diverse qualities of the substances resulting from their union, alike defy explanation by a non-theistic theory of evolution.

The origin of life, the first thrill of sensation, and the dawn of conscious intelligence, as well as the first perception of good and evil, not only remain as inscrutable as ever, but the many recent attempts made to obtain an explanation of them have only served to bring out more conspicuously their profound inscrutability. The ultimate constitution of matter, the origin and nature of vitality, feeling, and consciousness, as well as the intimate processes of life, growth, and reproduction, all these causes of the origin of species will, we believe, persistently remain quite inexplicable, though science will ever be fruitfully employed in elucidating more and more the means and processes of vital activity and organic change. The mystery of instinct, in spite of all the efforts of Darwin and his disciples, remains absolutely insoluble, and, instead of being capable of explanation by any other organic actions, vital activities — such as those of growth, repair of injuries, reproduction of lost parts, the development of the embryo, and new kinds of animal life — are rather to be regarded as essentially belonging to its own category.

The repair of its broken net by the spider, the instinctive casting off by a crab of a mutilated limb just at that point whence its renovation can take place, the arrangement by an insect of conditions suitable not for its own life but for the welfare of a progeny it will never see — each and all belong fundamentally to the same group of activities as do processes of organic repair and embryonic development. He who can fully understand instinct would (as we pointed out in the pages of this Review in the year 1881) possess a key capable of unlocking all the mysteries of organic nature.

We may now in conclusion briefly state, by way of resume, what we deem the essential facts of the "evolution of evolution", what was the origin, development of that hypothesis, what is its present state, and what we deem will be its fate in a more or less distant future.

"Evolution" as popularly understood, the evolution of "naturalism", of Spencer and of Darwin — the non-theistic system of evolution — was born in original sin, since its very name is a fraud, nay, rather a robbery! Its name gives the lie to its own conception, while it truly harmonizes with its very contradictory, namely, that system of theistic evolution to which we are confident it will sooner or later give place. For no process of "evolution" is possible unless there has been an antecedent process of "involution". The development of a germ, of a banyan tree from a seed; of a whole from a minute ovum; of a complex political organization from mere tribal customs; of all the arts from their veriest rudiments, and of the perfected "love" of God from initial tremors of his "fear"; can one and all be readily accepted as credible and reasonable because in each and every case the ultimate outcome existed *in potentia* from the first.

But what thirst can be quenched from an empty cup? What nourishment for mankind is to be raised from a field into which no seed has been ever cast?

The irrationality of its very name presided over its birth and origin, since it was begotten by junction of a system which did not even aspire to anything beyond sensation and imagination, with another absolutely suicidal, since it rejected that rational foundation upon which alone every logical system must be based.

But though here, as at the outset of this article, we lament the destruction of scholasticism and the advent of the sensationalism of Locke and the follies of Bacon and Descartes, we should be very sorry to be taken for a mere reactionary, whether in science, politics or religion. All attempts at restorations have invariably been failures, and as we have thus every reason to deem them impossible, we may well also believe that they are contrary to God's will.

Yet, although scholasticism can never be revived, its essential truths may be successfully advocated after having been assimilated by the modern mind and reproduced in modes possible for modern acceptance, as the Constitution of the United States enshrines the political principles maintained by St. Thomas Aquinas.

As we have depicted, the decay of pure philosophy coinciding with the rise of merely physical science, prepared the way for the theory of non-theistic evolution, and rendered its conception, birth and nutrition possible.

Culminating with the advent of Darwinism, it has now reigned over the modern scientific world —not without noteworthy protests — for a full generation.

Its present state, however, is one of incipient decay. It is honeycombed with doubts and difficulties, and the number of those who recognize that our ethical intentions alone are, as we long ago pointed out, absolutely fatal to it, is ever on the increase. But fresh discoveries in embryology, the repeated cutting down of phylogenetic trees, new views (such as those we have referred to) about the nature of cells and cell-structure are continually giving rise to conceptions which, from the Darwinian point of view, are heresies, though the heretics may be (and, as yet, generally are) the adherents of the non-rational theory of evolution.

But physiological progress is notably forming a fresh soil suitable for the growth of very different philosophical views from those which belong to "Naturalism". Notably, a new vitalism has been evolved by the recognition of the essential peculiarity of the activities of living organisms, for which mere physics and chemistry have been shown as utterly unable to account. Similarly, it is becoming more and more obvious and undeniable that a living creature is a unity — a unity of force no less than of bodily form. The existence of some immaterial principle of individuation in each animal and plant is more and more plainly indicated, and thus the truth of Aristotle's doctrine on the subject[35] (which we believe will never be bettered) is becoming more probable and more fully justified.

But these various influences — various in themselves, but alike hostile to naturalism — will be far surpassed in their efforts when, by slow degrees, the fundamental truths of philosophy — alike the prolegomena of all science and of all religion — have forced their way into the active, enterprising minds of a younger generation. Then the non-theistic theory of "evolution" (absurdly named evolution) will, by degrees, give place to a

---

35 For Aristotle's doctrine on the subject, see: Kraut, Richard, "Aristotle's Ethics", The Stanford Encyclopedia of Philosophy (Fall 2022 Edition), Edward N. Zalta & Uri Nodelman (eds.), URL = <https://plato.stanford.edu/archives/fall2022/entries/aristotle-ethics/>.

true, lexical system of evolution which will justify the name it bears.

According to that system, the whole universe was created by a Being the prototype of all wisdom, all virtue and all beauty, who bestowed on it the power to evolve, by His ever-present sustentation and concurrence, whatever He had involved within it at the first moment of its creation, and had thus truly Himself created *in potentia*. Thus all is order and harmony in the organic as well as in the inorganic world — life, feeling, thought, society, ethics, politics, art and science successively appearing with no less spontaneity than the various senses, instincts and habits of living but non-rational nature. To the knowledge which is merely direct and immediate succeeds knowledge which is more and more reflective, self-conscious and deliberate, till, under and through Divine inter-position — no less omnipotent because imperceptible — the highest activities of which human nature is capable are attained. Since the universe, as being essentially one, could never have been submitted to the action of any sort of natural selection, its power and properties must have been due to the creative will of God, and the results of their action (the perverse results of man's free-will excepted) must have been likewise preordained. Our knowledge is His gift, and our most important knowledge is that of our true relationship to Him, and thus what is at once the most important agent, as well as the highest end, of evolution, as rationally understood, is and must be, that which reveals to us our duties and our privileges — namely, religion.